"Bring me a pink drink!"
said the king.
He drank.

1

"Sing me a long song!"

he said.

His men sang.

"Do this!" said the king.
"Go there and do that!"

The men did lots of things.

Their king did not thank them.

Some of the men felt sad.

"I can think of a plan,"

said Frank.

"Come and help me!"
said the king.

"Are you going to thank
them?" said Frank.
"It is a thrill to get a
thank you from the king."

"I can thank them,"
said the king.
Then his men felt glad to help.

The End